目　次

1、回転動力軸装置の使用方法--2

2、回転動力軸装置の使用方法（英文解説）----------------------------7

3、公報解説　回転動力軸装置・実用新案登録第３１６８５５４号------9

　(1)　要約・課題--9

　(2)　解決手段--9

　(3)　実用新案登録請求の範囲----------------------------------9

　(4)　考案の詳細な説明--11

　(5)　技術分野--11

　(6)　背景技術--11

　(7)　考案の概要・考案が解決しようとする課題------------------12

　(8)　課題を解決するための手段------------------------------------12

　(9)　考案の効果--14

　(10)　図面の簡単な説明--14

　(11)　考案を実施するための形態------------------------------------15

　(12)　産業上の利用可能性--24

　(13)　符号の説明--24

4、英文--28

１、回転動力軸装置の使用方法

　今年ヨーロッパで確認されたあらゆる粒子を結びつける宇宙に満遍なく広がっている最後の粒子といわれるヒッグス粒子の大発見があった。　質量をもたらすこのヒッグス粒子は全ての空間の力場に影響を及ぼしていると思われる。　分子と分子との間に働く力を分子間力という。

　一般に分子はあまり近づくと反発する力を及ぼし合うが少し離れたときは互いに引力を及ぼし合う。　この反発力は交換反発力とクローン反発力とによって生じ引力は、ファン＝デル＝ワールス力（主として分散力）によって生ずる。　分子間のポテンシャルとしては$-u/r^m + v/r^n$で表わされる。　ｒは分子間隔・vは、それぞれの物質によって決まる定数でｍは普通６．ｎは８～１２を示し、前項が引力後項が反発力となっている。　r^nで示されるように分子間の距離が小さくなればなるほど反発力は急激に増大することになる。

　二つの分子間の距離が小さくなるためには分子と分子が近づく、従って、その反発力を押して引力が次第に増大することになる。　そこには微小的距離の引力の加速力が永久的に発生していることになる。

　ある位置で二つ或いは複数の分子は釣り合っているがその分子の運動が激しく高温になればバランスは崩れる。　地球のコアで半永久的に発生している高熱膨張エネルギーとそれを押さえ込もうとする引力（重力）エネルギーが釣り合っているのと同様なものである。　人は主に太陽のエネルギーで生活しているが、現在一般に言われていることはエネルギー保存の法則に反して低級になった熱は最終には広大な宇宙へ逃げて全て消滅するとされている。

　上空に行けば行くほど空気も薄くなり、分子の運動も弱まって気温も下がる。重力も影響して、全ての熱が無限大な宇宙へ逃げて消滅するのが原因だけではない

ようだ。　例えば地球内核は想像を絶するマグマ分子の運動が激しいところであるが地表面に近づくに従い分子間引力によって段々と運動量が減少して冷たい固体の岩石になって行く。

　分子の運動量が大きい高圧高温のマグマの熱エネルギは地上を離れ宇宙へ逃げて消滅するまでもなく大部分の熱は地殻部分の岩石分子の運動量が下がることによって消耗しているものと思われる。　マントルの灼熱のマグマと永久凍土との間で何が起きているのか近い内にヒッグス粒子の存在がその答えを導き出すだろう。

　太陽と同様に地球内核で継続して発生した高熱エネルギーは地表に届くまでに途中で消滅するのではなく分子から成り立った岩石等あらゆる物質物体内に備わっている分子間力によって取り込められているとしたなら、その分子間力を利用して、そのエネルギーを取り出し新しく生み出すことが可能である。　物体内（主として形状弾性体）に取り込まれて継続して生み出されたている分子間力として存在するエネルギーを有効な手段で取り出すことが可能である。

　生物にとって有害な放射能を大量に出す原子力は今後一切利用してはならない。そのような巨大な熱を発生するハイリクスな原子力より出力は小さくても安全な分子間力を利用する方法を用いて送電線の要らない発電をすれば、これ以上に自然環境を破壊する必要はない。

　分子間の引力と反発力を応用する方法として形状弾性のスプリングやバネを用いる。　外力によって、歪み変形を生じている物体が力を除かれたときに戻ろうとする力が弾性であるが弾性体の組み合わせ方法は幾つでもあり、無限に分子間力の効力を引き出させる。　この考案の回転動力軸装置は数ある弾性体の組み合わせ方法の中の一例である。

　分子間力は不増不減エネルギーであって、回転動力軸装置から取り出したエネル

ギーは最終には低級な熱になって再び空気や海水や岩石を構成している物質物体内の分子間力によって、取り込まれて公害の出ない自然を利用する再生可能エネルギーに似たものと定義づけられる。　大規模にこの技術が一般的に導入されると発電所からの電力供給量を激減させて地上の送電線は全て撤廃される。　発電所で使用する化石燃料は減り、経費や CO_2 の排出も抑えられる。　自動車もこの回転動力軸装置で故障も少なく、安全に動かせるし、電気自動車にすれば騒音問題も解消して排気ガスも吸わなくても良くなって健康長寿のために寄与できる。　又この技術から得られる回転力を動力源としているため通常の電気自動車に必要な充電も不要で走行距離に限界がない。　大電力を必要とする電車もこの装置を大規模にすればローコストで架線も不要である。　やがて航空機も給油が必要でない技術が開発されどの場所でも自由に飛んで行けて救助活動にも大きく貢献でき、ガス欠で墜落することはない。　無償のエネルギーで物が作れるため全ての物を安く手に入れることが出来る。　エネルギーが無尽蔵のため究極のリサイクル技術が確立されて有限な資源の消費を抑えて海水を真水にするコストも下げることが可能となって食糧の増産が見込め世界的人口増加で行き詰まった諸難題も解決できる。　現在の主に化石燃料使用の発電方法では発電段階で約 55％もの電力のロスがあり更に送電と変電で 20％のエネルギーを損失して家庭に届くまでに 75％の電力を失い低効率で無駄に資源を消費して取り返しのできない自然環境を破壊する。

　究極的には完全な分散型電源を実現して発電所を廃止する。　これで危険な放射性廃棄物を出す原子力発電所も役割は終焉する。　この技術は人類全体に化石燃料の使用を放棄させる発端となりエネルギー問題を完全に解決する。　この技術が確立され普及するとあらゆる方面に良い波及効果を及ぼし人類は飛躍的に進化する。

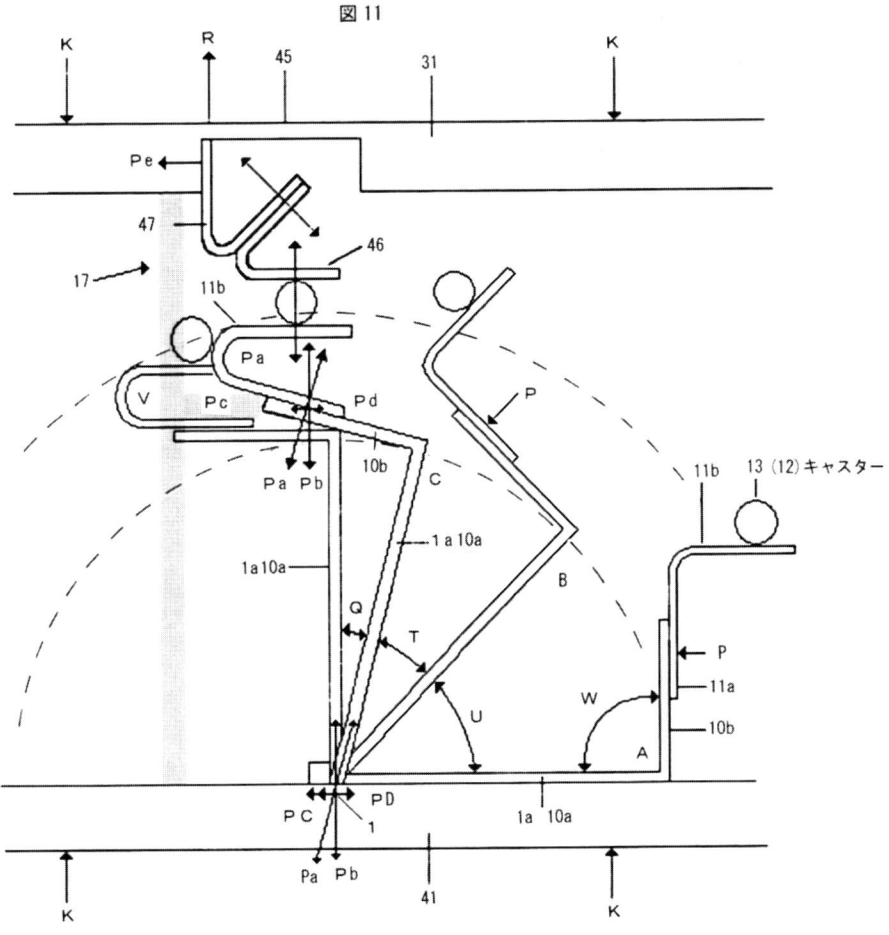

図11

　この回転動力装置の安定回転のために新たにストッパーVを取り付けることを提案したいと思います。　従来の製品は出来上がった部分をそのまま組み合わせ加工すればよかったけれどもこの装置は弾性体部品に外力を加えながら安定した状態を保ちつつ組み立てなければならないため試作品の完成段階において諸経費がかかるものですが、一度完成すれば安価に大量生産が可能で故障もしません。

　図面10.11において外力をPをかける直前の段階AからBを経過して最始動時のCとなる。　そのC点において中間部材10の腕片部10bの先端部へ上下の円形鋼版31、41へ新規に固定した止め具ストッパーVを接触させる。　上下の回転動力軸30、40が回転中突然制動力をかけた時に上下円形鋼版31、41に固

5

定されていない弾性体結合部品が慣性によってすべて前のめりとなり調和している組み合った弾性体のバランスを崩すのを制止する。

　上板バネ４７下板バネ４６結合体の連続機構１７を取り外した場合にもストッパーＶがあると同様にして弧を描いて前のめり状態にならずに円形鋼版３１，４１の間では前進後退も出来ずにＣからＢＡと返ることはない。

　この装置から発生した運動エネルギーは他の熱機関でも同様だがあらゆる仕事をした結果、低級な熱となって最終には分子の運動量が低下した状態で空気や水、物質物内に取り込められる。

　分子間力が激しい状態が高温状態であり、熱エネルギーはすべて宇宙へ逃げて消滅したりはしない。　地球内核で発生した高熱は地殻部で分子間力によって段々と低下して地表近くでは宇宙へ逃げるまでもなく下がっている。　再生可能エネルギーは弾性体の分子間力と反発力を利用すれば経済効果が高い。また不十分な所が多々あると思いますが経済振興のために長期未来の視野に立ち是非ともご検討をください。

板バネに外力をかける前段階の状態図　　セットを完了してバネを開放する状態図

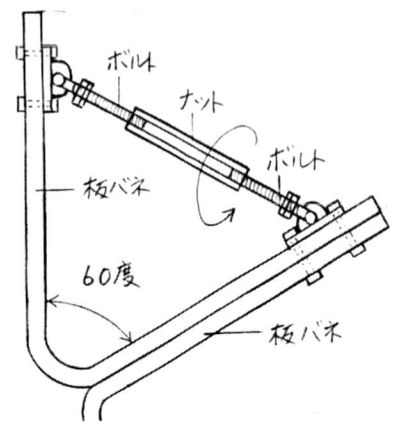
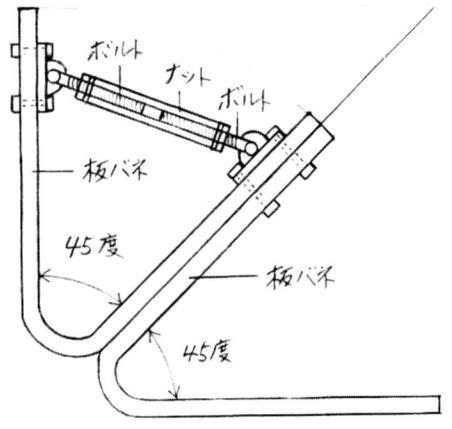

2、The usage of rotation power axis equipment

There was large discovery of the Higgs boson called particles of the last uniformly spreading in the universe which ties up all the particles checked in Europe this year.

The power committed between the molecule and molecule which are considered that this Higgs boson has affected the power place of all the space is called intermolecular force.

The more it goes high up in the sky, air also becomes thin, movement of a molecule also becomes weaker, and, the more temperature also falls.

It does not seem to be only because gravity also influences and all the heat escapes and disappears to the infinite universe.

For example, with an intermolecular attractive force, steps and quantity of motion decrease, and it becomes the rock of a cold solid and goes as it approaches earth surface, although an earth's inner core is a place where movement of the magma molecule which is beyond imagination is intense.

Balance is the same as the high-fever expansion energy semipermanently generated with the core of the collapsing earth and the gravitation (gravity) energy which tries to press down and hold it down balance if movement of two or two or more molecules becomes high temperature violently in a certain position.

Although people mainly live on solar energy, it is supposed at last that it escapes to the vast universe and disappears altogether the heat with which saying to the general present became low-grade against the law of conservation of energy.

The more it goes high up in the sky, air also becomes thin, movement of a molecule also becomes weaker, and, the more temperature also falls.

It does not seem to be only because gravity also influences and all the heat escapes and disappears to the infinite universe.

For example, with an intermolecular attractive force, steps and quantity of motion decrease, and it becomes the rock of a cold solid and goes as it approaches earth surface, although an earth's inner core is a place where movement of the magma molecule which is beyond imagination is intense.

The heat energy of the magma of high-pressure high temperature with large quantity of motion of a molecule does not need to leave the ground, it is not necessary to escape and disappear to the universe, and it seems that most heat disappears when the quantity of motion of the rock molecule of an earth portion falls.

Existence of a Higgs boson will draw the answer for what has occurred between the magma of red heat of a mantle, and a permafrost layer, while it is near.

 A perfect dispersed-type power source is realized ultimately, and plant is abolished.

A role carries out the end also of the nuclear power plant from which dangerous radioactive waste is taken out with this.

This technology serves as the beginning which makes the whole body of mankind abandon use of a fossil fuel, and solves an energy problem completely. If this technology is established and it spreads, a ripple effect good for all directions will be done, and human beings will evolve by leaps and bounds.

3、公報解説　回転動力軸装置・実用新案登録第３１６８５５４号

実用新案登録第３１６８５５４号
考案の名称；回転動力軸装置
実用新案権者；奥原　順応

⑴【要約】【課題】
　最初に外力を加えれば、長時間回転することの出来る動力装置を提供する。

⑵【解決手段】　下回転動力軸４０の上端部に下円形鋼板４１を固定し、下円形鋼板の上に凹部か、２つの偶力受止突起１５を設け、交差角を有する両端腕部を有するスプリング１の中央胴部を両突起に係止し、Ｌ字状をした中間部材１０の基片部１０ａをスプリングの腕部１ａ，１ｂに嵌合固定し、中間部材の腕片部１０ｂにＬ形バネ１１の下部分１１ａを固着し、中間部材の腕片部に取付けたところのバネにキャスター１２を介して鋼球ボール１３を配置し、下円形鋼板の外周縁近くに立設した数本の鋼鉄支柱２を植設固着し、鋼鉄支柱の上部に上円形鋼板３１の縦孔４３を嵌め、スプリングの両腕部を起こして鋼球ボールを上円形鋼板の下面に当接させ、上円形鋼板上に上回転動力軸３０を固着立設したものである。
【選択図】　図１

⑶【実用新案登録請求の範囲】
【請求項１】
制動装置（４４）に挿通した下回転動力軸（４０）の上端部に下円形鋼板（４１）を固定し、交差角を有する両端腕部（１ａ，１ｂ）を有する主スプリング（１）の中央胴部を下円形鋼板（４１）係止機構（１５）を介して係止し、二つのＬ字状中間部材（１０，１０）の基片部（１０ａ）を主スプリング（１）の両腕部にそれぞれ固定し、中間部材（１０）の腕片部（１０ｂ）に二つのＬ形スプリング（１１）の下部分（１１ａ）を取付け、遠心力止め連結部材（９）を両腕片部（１０ｂ，１０ｂ）の間に架設し、下円形鋼板（４１）に少なくとも三本の鋼鉄支柱（２）を植設し、鋼鉄支柱（２）の上部に上円形鋼板（３１）の縦孔（４３）を嵌めて固定し、Ｌ形スプリングの上片部（１１ｂ）を上円形鋼板の下側に連接機構（１７）を介して配置し、上円形鋼板上に上回転動力軸（３０）を固着立設した回転動力軸装置。
【請求項２】
制動装置（４４）及び下軸受（４２）に挿通支承した下回転動力軸（４０）の上端

部に下円形鋼板（４１）を固定し、交差角を有する両端腕部（１ａ，１ｂ）を有する主スプリング（１）の中央胴部を下円形鋼板（４１）上に係止機構（１５）を介して係止し、二つのＬ字状をした中間部材（１０，１０）の基片部（１０ａ）を主スプリング（１）の両腕部にそれぞれ嵌合固定し、中間部材（１０）の腕片部（１０ｂ）に二つのＬ形スプリング（１１）の下部分（１１ａ）をスライド位置調節可能に取付け、伸縮可能な遠心力止め連結部材（９）を両腕片部（１０ｂ，１０ｂ）の間に架設し、下円形鋼板（４１）の外周縁近くに立設した少なくとも三本の鋼鉄支柱（２）を植設固着し、鋼鉄支柱（２）の上部に切ったネジ部（２ａ）に位置決めナット（３）を螺合して上円形鋼板（３１）の縦孔（４３）に嵌め、更に締付ナット（４）を螺合し、Ｌ形スプリング（１１）の上辺腕部（１１ｂ）を上円形鋼板の下側に連接機構（１７）を介して配置し、上円形鋼板上に上回転動力軸（３０）を固着立設し、上回転動力軸に対し軸方向位置調整可能な止軸受（３２）を配置した回転動力軸装置。

【請求項３】
二つの突起（１５ａ，１５ａ）により係止機構（１５）を構成した請求項１又は２に記載の回転動力軸装置。

【請求項４】
凹溝（１５ｂ）により係止機構（１５）を構成した請求項１又は２に記載の回転動力軸装置。

【請求項５】
上辺部（１１ｂ）にキャスター（１２）を介して鋼球ボール（１３）を配して連接機構（１７）を構成した請求項１乃至４のいずれかに記載の回転動力軸装置。

【請求項６】
上円形鋼板（３１）の下面に設けた凹部（４５）にＬ型の上板バネ（４７）の垂直片を取付け、上板バネ（４７）に取付けたＬ型の下板バネ（４６）の上辺を上板バネ（４７）の水平片に取付け、下板バネ（４６）の水平辺をＬ形スプリング（１１ｂ）に取付けた請求項１または２のいずれかに記載の回転動力軸装置。

【請求項７】
下板バネ（４６）の水平辺を鋼球ボール（１３）上に接触させた請求項５に記載の回転動力軸装置。

【請求項８】
制動装置（４４）及び下軸受（４２）に挿通支承した下回転動力軸（４０）の上端部に下円形鋼板（４１）を固定し、下円形鋼板（４１）の上にスプリング（１）を嵌る凹部か、或は２つの偶力受止突起（１５，１５）を設け、交差角を有する両端

腕部（１ａ，１ｂ）を有するスプリング（１）の中央胴部を突起（１５，１５）に係止し、Ｌ字状をした中間部材（１０，１０）の基片部（１０ａ）をスプリング（１）の腕部に嵌合固定し、中間部材（１０）の腕片部（１０ｂ）にＬ形バネ（１１）の下部分（１１ａ）を固着し、伸縮可能な遠心力止め連結部材（９）を両中間部材の間に架設し、下部分（１１ａ）をスライド位置調節可能に中間部材の腕片部（１０ｂ）に取付けたところのバネ（１１，１１）の上辺部（１１ｂ）にキャスター（１２，１２）を介して鋼球ボール（１３，１３）を配置し、下円形鋼板（４１）の外周縁近くに立設した数本の鋼鉄支柱（２）を植設固着し、鋼鉄支柱（２）の上部に切ったネジ部（２ａ）に位置決めナット（３）を螺合して上円形鋼板（３１）の縦孔（４３）に嵌め、更に締付ナット（４）を螺合し、スプリングの両腕部を起こして鋼球ボールを上円形鋼板の下面に当接させ、上円形鋼板上に上回転動力軸（３０）を固着立設し、上回転動力軸に対し軸方向位置調整可能な止軸受（３２）を配置した回転動力軸装置。

(4)【考案の詳細な説明】
(5)【技術分野】
【０００１】
この考案は、最初に外力を加えておくと、その外力を取除いても長時間回転することの出来る回転動力軸装置に関するものである。
(6)【背景技術】
【０００２】
最初に外力を加えれば、その外力を取除いても回転する装置として、例えばぜんまい式時計があげられる。
【０００３】
物質を構成する微粒子を分子とよんでいる。分子は、温度が下がれば、分子の運動はより強く制限されて固体となる。しかし、その場合でも、独立した分子は存在するが、分子同士の間では強い化学結合力は認められず、分子が凝集しているのは、化学結合力とくらべるといちじるしく弱い力のファン・デル・ワールス力によることが多い。したがって、このような場合も分子からなる物質である。分子はあまり近づくと反発する力を及ぼし合うが、少し離れたときは互いに引力を及ぼし合う。分子間の距離が小さくなればなるほど反発力は急激に増大することになる。分子間は化学結合によらず、分子間力で結ばれているので、その力はきわめてゆるい。
【０００４】
分子間の距離が小さくなればなるほど反発力は急激に増大することになるとある

が、二つの分子間の距離が小さくなるためには分子と分子が近づく。したがって、その反発力を押して、引力が次第に増大することになる。そこには、微小的距離の引力の加速力が永久に発生していることになる。そしてある位置で二つ或いは複数の分子の引力と反発力が釣り合って静止状態に見えるだけである。地球内部の熱膨張エネルギーと引力（重力）エネルギーが釣り合っているのと同じようなものである。
【０００５】
弾性には、体積の変化に対しておこる体積弾性と、形の変化に対しておこる形状弾性がある。ばねの弾力などは、おもに形状弾性によっておこる。固体では形状弾性と体積弾性がともにおこり、両方がいっしょになって現われる場合が多い。棒や板を曲げると、板の外側の面は伸びて内側の面は縮むが、その中間の面は曲がるだけで伸縮がない。棒や針金の一端を固定して、他端に偶力を加えてねじった場合のねじれ変形で、断面が円でない場合には、ねじり剛性は複雑である。また、棒の中の多数の微小粒子間の作用と反作用とが抵抗している。

(7)【考案の概要】【考案が解決しようとする課題】
【０００６】
従来のぜんまい式のもの等に比べて、長時間回転することの出来る動力装置を提供しようとするものである。
(8)【課題を解決するための手段】
【０００７】
図面を参考にして説明する。
請求項１の考案は、制動装置４４に挿通した下回転動力軸４０の上端部に下円形鋼板４１を固定し、交差角を有する両端腕部１ａ，１ｂを有する主スプリング１の中央胴部を下円形鋼板４１係止機構１５を介して係止し、二つのＬ字状中間部材１０，１０の基片部１０ａを主スプリング１の両腕部にそれぞれ固定し、中間部材１０の腕片部１０ｂに二つのＬ形スプリング１１の下部分１１ａを取付け、遠心力止め連結部材９を両腕片部１０ｂ，１０ｂの間に架設し、下円形鋼板４１に少なくとも三本の鋼鉄支柱２を植設し、鋼鉄支柱２の上部に上円形鋼板３１の縦孔４３を嵌めて固定し、Ｌ形スプリングの上片部１１ｂを上円形鋼板の下側に連接機構１７を介して配置し、上円形鋼板上に上回転動力軸３０を固着立設したものである。
【０００８】
請求項２の考案は、制動装置４４及び下軸受４２に挿通支承した下回転動力軸４０の上端部に下円形鋼板４１を固定し、交差角を有する両端腕部１ａ，１ｂを有する

主スプリング１の中央胴部を下円形鋼板４１上に係止機構１５を介して係止し、二つのＬ字状をした中間部材１０，１０の基片部１０ａを主スプリング１の両腕部にそれぞれ嵌合固定し、中間部材１０の腕片部１０ｂに二つのＬ形スプリング１１の下部分１１ａをスライド位置調節可能に取付け、伸縮可能な遠心力止め連結部材９を両腕片部１０ｂ，１０ｂの間に架設し、下円形鋼板４１の外周縁近くに立設した少なくとも三本の鋼鉄支柱２を植設固着し、鋼鉄支柱２の上部に切ったネジ部２ａに位置決めナット３を螺合して上円形鋼板３１の縦孔４３に嵌め、更に締付ナット４を螺合し、Ｌ形スプリング１１の上辺腕部１１ｂを上円形鋼板の下側に連接機構１７を介して配置し、上円形鋼板上に上回転動力軸３０を固着立設し、上回転動力軸に対し軸方向位置調整可能な止軸受３２を配置したものである。
【０００９】
請求項３の考案は、請求項１または２の考案において、二つの突起１５ａ，１５ａにより係止機構１５を構成したものである。
【００１０】
請求項４の考案は、請求項１または２の考案において、凹溝１５ｂにより係止機構１５を構成したものである。
【００１１】
請求項５の考案は、請求項１乃至４の考案において、上辺部１１ｂにキャスター１２を介して鋼球ボール１３を配して連接機構１７を構成したものである。
【００１２】
請求項６の考案は、請求項１又は２の考案において、上円形鋼板３１の下面に設けた凹部４５にＬ型の上板バネ４７の垂直片を取付け、上板バネ４７に取付けたＬ型の下板バネ４６の上辺を上板バネ４７の水平片に取付け、下板バネ４６の水平辺をＬ形スプリング１１ｂに取付けたものである。
【００１３】
請求項７の考案は、請求項５の考案において、下板バネ４６の水平辺を鋼球ボール１３上に接触させたものである。
【００１４】
請求項８の考案は、制動装置４４及びド軸受４２に挿通支承した下回転動力軸４０の上端部に下円形鋼板４１を固定し、下円形鋼板４１の上にスプリング１を嵌る凹部か、或は２つの偶力受止突起１５，１５を設け、交差角を有する両端腕部１ａ，１ｂを有するスプリング１の中央胴部を突起１５，１５に係止し、Ｌ字状をした中間部材１０，１０の基片部１０ａをスプリング１の腕部に嵌合固定し、中間部材１０の腕片部１０ｂにＬ形バネ１１の下部分１１ａを固着し、伸縮可能な遠心力止め

連結部材9を両中間部材の間に架設し、下部分11aをスライド位置調節可能に中間部材の腕片部10bに取付けたところのバネ11,11の上辺部11bにキャスター12,12を介して鋼球ボール13,13を配置し、下円形鋼板41の外周縁近くに立設した数本の鋼鉄支柱2を植設固着し、鋼鉄支柱2の上部に切ったネジ部2aに位置決めナット3を螺合して上円形鋼板31の縦孔43に嵌め、更に締付ナット4を螺合し、スプリングの両腕部を起こして鋼球ボールを上円形鋼板の下面に当接させ、上円形鋼板上に上回転動力軸30を固着立設し、上回転動力軸に対し軸方向位置調整可能な止軸受32を配置したものである。

(9)【考案の効果】
【0015】
スプリング1の両端の腕部1a,1bが上部で固定した中間部材10,10の上部分に固定されたバネ11,11の表面に直角にかけ続けているところの偶力の外力を解除しても上方から降下する上円形鋼板31から2箇所のキャスター12,12を挟んでのバネ11,11の上部分にかかる偶力ではない圧縮外力のはたらきでバネ11,11の下部分（外力の偶力を最初から継続してかけ続けられていた部分）11aは押され元の状態に応力で返ろうとするが返れない。

【0016】
弾性体に外力が作用するときその外力に抵抗して弾性体に抵抗して弾性体にそのままの形を保とうとして生ずる内力（応力）は反作用の力である。応力は分子の組成が高熱とかで変化しない限り半永久的に続く。分子はあまり近づくと反発する力を及ぼし合うが少し離れたときには互いに分子間の引力を半永久的に及ぼし合う。その結果バネ11,11の下部分（スプリング1の両端の腕部1a,1bの上部分で固定されたL字形に曲げられている中間部材10,10の上部）に偶力を新たに生み出す。

【0017】
制動装置44のブレーキが解放され上下回転動力軸30,40が回転を始める状態の図2と図3のCの状態にあるように上下の円形鋼板31,41からの圧縮外力以外の外力（偶力）を解消すれば、スプリング1の下部は元の状態に応力弾性力で返ろうとするために、スプリング1のド部全体を嵌合した下円形鋼板41の中央部分凹部で生じる効果と同じ偶力受止突起15,15に対し偶力をそして上下の円形鋼板31,41と固定され一体化した上下の回転動力軸30,40に半永久的に回転力を発生する。

(10)【図面の簡単な説明】
【0018】

【図１】装置の初期状態の一部分解斜視図である。
【図２】装置全体の斜視図である。
【図３】バネの作動説明図である。
【図４】一部を変形した要部の鉛直切断拡大正面図である。
【図５】係止機構を凹溝にしたときの実施形態図である。（イ）は垂直切断面図で、（ロ）のイーイ断面図、（ロ）は平面図で、（イ）のローロ断面図である。
【図６】連接機構の第２例の初期図である。
【図７】図６の中間図である。
【図８】図６の最終図である。
【図９】図６の異なる最終図である。
【図１０】主スプリングとＬ形スプリングの作動説明図である。
【図１１】異なる主スプリングとＬ型スプリングの作動説明図である。
【図１２】図１の一部分解図である。
【図１２】図１０の装置の初期状態の一部分解斜視図である。
【図１３】図２の一部分解図である。
【図１３】図１０の装置全体の斜視図である。

(11)【考案を実施するための形態】
【００１９】
図を参考にして説明する。図１，図２及び図３の行程段階Ａ．Ｂ．Ｃの状態では、下回転動力軸４０の回転力を制御する制動装置４４はまだ作動中の状態であり、下回転動力軸４０と下円形鋼板４１は固定され、その下円形鋼板４１に固定された左右２箇所の偶力受止突起１５，１５もしくは下円形鋼板上にスプリング１がすれすれか、僅かに残る程度に嵌合させるスプリング１の下部分全体を嵌める凹部を設ける。これに対してそれに接触するスプリング１の中央胴部の両端へ偶力の外部からの外力Ｐをかける。当然ながら、行程段階ＡよりもＢの段階、更にＣの段階へとこれを組み立てるときに外部からの偶力の外力は大きい力を順次かけ続けなければならない。
【００２０】
同様同時にスプリング１の両端の腕部１ａ，１ｂに被せたＬ字形の鋼管１０ａと固定した鋼鉄棒１０ｂよりなる中間部材１０，１０の上部に固定されたバネ１１，１１の下部分である基片部１１ａの表面上に直角し、且つもう一方の上片部１１ｂと平行な偶力の外力Ｐを図３のＣの状態になるまでかけ続ける。そして最終は図２のようになる。但し、その場合、図３の行程段階Ｂの時点での偶力の外力Ｐに加勢す

ることになる上下の円形鋼板３１，４１を近付ける圧縮圧力の回転動力軸３０，４０に平行する外部からの垂直方向の外力を図３のＣの状態になるまで一緒に同時加えてはならない。一応、混乱することを防ぐための説明のためにそのようにする。

外力Ｐは順次Ｃの段階にきてから終わった時点で、外力が大きくなってかかる偶力の外力Ｐと同一の力と力の効果をもたらすまったく別な外力を受けて図２と図３のＣのように変形したバネ１１ａ，１１ｂ，１１ａ，１１ｂの弾性力のＰａと、その反力Ｒの外力を上下の円形鋼板３１，４１とバネ１１ｂ，１１ｂが平行するようにバネ１１ｂ，１１ｂの両方の表面にかけて組み込んだ後に、その位置で次に代って上下の円形鋼板３１，４１を近付けて圧縮の外力をバネ１１ｂ，１１ｂの表面上に円形鋼板３１，４１とバネ１１ｂとが平行するようにかける。だから、段階ごとに外力を解消すると同時に、それとまったく同じ効力のある新しい外力を順次かける。

図３の行程段階Ａはバネ腕部１ａ基片部１０ａに対し、垂直方向から見れば９０度ということである。ＢはＡの状態から外力Ｐを受けて４５度ほど立った状態で、Ｃは更に外力Ｐを受けてＢの状態から２３度起きた状態であり、Ｃの位置での角度に限定はせず、その前後の角度であればよい。

【００２１】
図１の上方にある中心に上回転動力軸３０を固定した上円形鋼板３１をそれに平行した下部の下円形鋼板４１を固定した４箇所の鋼鉄支柱２の一部にあるネジ部２ａとそれを締める上下のナット４，３によって間にある上円形鋼板３１を下方へ移動させて上下の円形鋼板３１，４１を近付ける。

【００２２】
上円形鋼板３１の４箇所は中空の縦孔４３となっており、その中を４本の鋼鉄支柱２の上部分のネジ部２ａが通る。

【００２３】
バネ１１の上片部１１ｂの平坦な表面上へ固定されたキャスター１２を挟んで上円形鋼板３１の上の締付ナット４を締めることによって徐々に下降させてバネ１１，１１の上方に固定されたキャスター１２，１２と平行或は水平な状態となるように下げて上方からの図３のＣの状態の位置での外力と同じ力の外力Ｐａをかけることとする。Ｒは反力（ただし、Ｐ＝Ｐａ＝Ｒ）、Ｐｂは外力Ｐａの垂直方向成分、Ｐｃは外力Ｐａの水平方向成分である。

【００２４】
平行する上下の円形鋼板３１，４１の間では、継続して下部の偶力受止突起１５，

１５へかけられた横方向の外力の偶力とは違う垂直の縦方向の圧縮圧力の外力をスプリング１及びバネ１１，１１に対してかけることにする。
【００２５】
図３でのＡの行程段階では制作時に曲げられたスプリング１の角度は９０度であるが対なので１８０度となる。
【００２６】
スプリング１の角度は制作時にそれが持っている弾性力の強弱によって、バネ１１の曲げられている角度と強度とに程良く釣り合わなければ上下の円形鋼板３１，４１に固定した上下の回転動力軸３０，４０の回転力を有効的に引き出すことは出来ない。
【００２７】
従って制作時にスプリング１の弾性体としての弾力が弱ければ対で行程段階Ａの１８０度以上、強ければ対で行程段階Ｂの９０度の角度でも良い。そして、同様にバネ１１も上方からの外力がかけられる前段階の図３のＡとＢの状態では９０度であるがそれ以上又以下の角度であっても良い。
【００２８】
スプリング１とバネ１１を一体化したものを上下の円形鋼板３１，４１内に組み入れるとき、あらかじめ曲げて製造したスプリング１とバネ１１の角度と弾性力が最適な状態で間に挟まれたキャスター１２が上下の円形鋼板３１，４１と平行な水平状態で保持される必要があり、どちらかが強くても或は弱くてもいけない。
【００２９】
図２と図３のようにＡから始まりＢ、そして最終段階のＣの状態になるとスプリング１の両端の腕部１ａ，１ｂが上部で固定した鋼鉄棒よりなる中間部材１０の上部分に固定されたバネ１１ａの表面に直角にかけ続けているところの偶力の外力を解除しても上方から降下する上円形鋼板３１から２箇所のキャスター１２，１２を挟んでのバネ１１，１１の上部分にかかる偶力ではない圧縮外力のはたらきでバネ１１の下部分（外力の偶力を最初から継続してかけ続けられていた部分）１１ａは押され元の状態に応力で返ろうとするが戻れない。
【００３０】
弾性体に外力が作用するときその外力に抵抗して弾性体に抵抗して弾性体にそのままの形を保とうとして生ずる内力（応力）は反作用の力である。応力は分子の組成が高熱とかで変化しない限り半永久的に続く。分子はあまり近づくと反発する力を及ぼし合うが少し離れたときには互いに分子間の引力を半永久的に及ぼし合う。
【００３１】

その結果バネ１１の下部分（スプリング１の両端の腕部１ａ，１ｂの上部分で固定されたＬ字形に曲げられている中間部材１０の上部）に偶力を新たに生み出すことになる。

【００３２】
制動装置４４のブレーキが解放され上下回転動力軸３０，４０が回転を始める状態の図３のＣ及び図２の状態にあるように上下の円形鋼板３１，４１からの圧縮外力以外の外力（偶力）を解消すれば、スプリング１の下部は元の状態に応力弾性力で返ろうとするために偶力受止突起１５，１５或はスプリング１の下部を嵌合した円形鋼板４１の中央凹部に対し偶力をそして上下の円形鋼板３１，４１と固定され一体化した上下の回転動力軸３０，４０に半永久的に回転力を発生する。

【００３３】
上下の円形鋼板４１，３１の中心部に上下の回転動力軸４０，３０を固定して上下の軸受４２，３２で支える。

【００３４】
スプリング１の上部分は始めの偶力の外力をかけた元の方向へ同じ力の偶力（弾性力応力）で返ろうとするが、円形鋼板３１，４１の上下からの縦方向の外力圧力によってスプリング１の上部分で固定されたバネ１１の下部は縦方向と元の外力をかけた方向へ返ろうとする弾性力の反対の新しく生じた同等の偶力を同時に受ける。

【００３５】
スプリング１の上部分に固定されたバネ１１の下部にかける偶力の外力を解消すると同時にそれとまったく同等の偶力をバネ１１の上部へ固定されたキャスター１２を間に挟み上下の円形鋼板３１，４１を狭めることによって同じはたらきをする外力の偶力をバネ１１，１１の下部分にかけることが出来る。

【００３６】
従って、スプリング１の下部（偶力受止突起１５，１５）で発生する偶力とスプリング１の上部（バネ１１，１１の下部）で発生する偶力は同じであるが、上部で発生する２つの相対した偶力は互いを打ち消し合う状態にする。

【００３７】
もしそこでバネ１１の上部はキャスター１２を固定したままスプリング１の下部の両端にかかる偶力の反対方向の偶力をもって偶力の外力をかける前の状態（図１のＡと図３のＡ，Ｂ）に弾性力で返ろうとしても間に圧縮圧力の外力を上下の円形鋼板３１，４１から受けたキャスター１２があるため絶対に返ることは出来ない。図３の行程段階Ｂは上部分でバネ１１ｂが二箇所程書いてあるが、これはＣからＢ

へ返ろうとしても上下の円形鋼板３１，４１があるためＣの状態からＢへスプリング１の部分を支点として、円運動で返ることは出来ないことを示すためのものである。

【００３８】
圧縮圧力の外力の作用反作用の方向は上下の回転動力軸３０，４０と平行しているので、その回転力の妨げにならない。

【００３９】
中間部材１０，１０のＬ字形に曲げた角度は上下の円形鋼板３１，４１の間で最も効率の良い回転動力軸３０，４０の回転力を生み出す角度が好ましい。図３では、９０度であるが、１３５度前後のへ字形からＬ字形の９０度の間で曲げた角度も良い。

【００４０】
スプリング１は図１，図２では棒バネだが、コイル状スプリングか板状スプリングでも可能である。バネ１１は図１，図２では板バネだが、コイルバネでも可能である。

【００４１】
図３の行程段階Ｂ，Ｃをみれば応力弾力で元に返ろうとしても上下の円形鋼板３１，４１に挟まれているために図２の４本の鋼鉄支柱２の上方の締付ナット４を緩めない限り、そして下部では偶力受止突起１５，１５で阻まれているので、前進や後退も出来ない。

【００４２】
しかしながら、スプリング１とバネ１１を一体化して独立した物体は静止でもそれを組み込んだ上下の円形鋼板３１，４１と回転動力軸３０，４０の本体は静止状態のままではなくスプリング１に偶力の外力をかけ始めた時点から回転動力軸３０，４０の回転力を制止した制動装置４４は継続して作動中なのでブレーキを解除すれば、偶力受止突起１５，１５では継続して回転力を受け続けているので上下の円形鋼板３１，４１に固定されている回転動力軸３０，４０はただちに回転出来る状態になる。

【００４３】
あらゆる物質物体を構成している分子は半永久的に反発力か引力を継続的に発生しているので、回転動力軸３０，４０に負荷がかからなければ加速力がつき回転数、回転力ともに高まるために一定置きに絶えず制動装置４４で制御しなければならない。

【００４４】

上下の円形鋼板３１，４１とバネ１１の上部とキャスター１２が平行な状態になるようにスプリング１とバネ１１の強度と角度を計算して製造することが必要である。
【００４５】
中間部材１０のＬ字形の上部の腕片部１０ｂとバネ１１の下部分１１ａを固定する部分はスプリング１とバネ１１を釣り合わせる微調節のためにスライド形式として板バネ１１の下部分１１ａに設けた長孔１１ｃを、中間部材１０の腕片部１０ｂに植設したスタッド１０ｃに通し、ナット１０ｄで締付けて中間部材１０にバネ１１を固定する。
【００４６】
また、図１，図２では下部分１１ａの下端縁に下に向かって延設配置したネジ棒１１ｅを腕片部１０ｂに設けた目玉具１０ｅに通しナット１１ｆで固定しているが、図４のものでは長孔１１ｃに２本のスタッド１０ｃ，１０ｃを通し、ナット１０ｄ，１０ｄで締付けている。
【００４７】
同様にバネ１１の上辺部１１ｂに設けた透孔１１ｄにキャスターケース１２ａに植設した２本のスタッド１２ｂを通し、ナット１２ｃで締付けてキャスターケース１２ａにバネ１１を固定する。
【００４８】
上円形鋼板３１の下面には、上鋼球７ａを載せた受皿３１ａが配置され、受皿３１ａを下から支持する保持具３１ｂが上円形鋼板３１の下面に固定されている。キャスターケース１２ａの上向き開口凹穴内には下鋼球７ｃを配してキャスター１２が抜止め不能に装着され、キャスター１２の上向き開口凹穴内には中鋼球７ｂを介して鋼球ボール１３が配置され、鋼球ボール１３は上受皿３１ａの下面に接触している。
【００４９】
バネ１１の上部の上片部１１ｂに固定されたキャスター１２の鋼球ボール１３は上下の円形鋼板３１，４１から受ける上下からの強力な圧力と摩擦力で上円形鋼板３１の裏面の表面上に接する一定位置に前進や後退も出来ずに保定され静止して円形鋼板４１の上の凹部４５か、偶力受止突起１５，１５が受ける偶力の作用横方向へ上下円形鋼板３１，４１と一体化して回転動力軸３０，４０は連動して継続回転する。
【００５０】
回転力以上の制動力が回転動力軸３０，４０にはたらけば、回転は止められる。

【００５１】
巨大な摩擦力を受けてもキャスター１２の内部にある鋼球ボール１３はスプリング１とバネ１１の釣り合いを調節するために一時だけ自由自在に回転出来て円形鋼板３１や回転動力軸３０，４０の回転力にマイナスの影響を及ぼさない。そのためには、キャスター１２の内部にある鋼球ボールは三重四重形式も必要である。
【００５２】
しかし、スプリング１とバネ１１を一体化したものを上下円形鋼板３１，４１の内側へ組み入れたときに、それが及ぼす回転動力軸３０，４０への継続した回転力にマイナスの影響がまったく生じないのならばむしろ摩擦力は必要なものでキャスター１２或は、遠心力止め螺筒１６と伸縮連結部材９は不要となりバネ１１の上部は直接にその上部にある円形鋼板３１の裏面に固定する。或は接するだけで同様に回転動力軸３０，４０を継続して回転させる。そのときには、遠心力止め螺筒１６と伸縮連結部材９は結合する。
【００５３】
中間部材１０の上部に固定されるバネ１１の下部分はスライド形式なので、上下の円形鋼板３１，４１とバネ１１，１１の上面が平行な水平状態をつくるためにてこの原理でスプリング１とバネ１１，１１の弾力の微調節をする。
【００５４】
最始動時に図式で下円形鋼板の凹部か偶力受止突起１５，１５は継続して偶力を受けたまま制動装置４４の制動力を解放する直前に遠心力止め螺筒１６と螺棒状の伸縮連携部材９を連結して回転運動中にスプリング１の両端部と左右のバネ１１，１１が遠心力で開いてしまわないようにする。開くとスプリング１とバネ１１，１１及び全体のバランスが崩れて損傷や回転動力軸３０，４０を止めることがあるので上下２箇所程必要なこともある。しかし、スプリング１とバネ１１の釣り合いの調節が済み再始動時に上円形鋼板３１の裏面とそれに平行するバネ１１ｂの上面を当接させボトルとナットで固定すれば遠心力止め螺筒１６と伸縮連結部材９は回転運動中に遠心力で開かないので不要である。
【００５５】
回転運動中は危険なので上下の軸受３２，４２と制動装置４４にカバー８を取付けて一体化してこの回転動力軸装置を設置する。
【００５６】
下部の一部が鋼管或は全体が鋼鉄棒の中間部材１０，１０のＬ字形に１３５度前後から９０度で曲がった所は上下からの圧力に弱いので製造時に補強片１４，１４によって補強する。スプリング１とバネ１１，１１は弾力の大きいものを使用すれば

21

回転動力軸３０，４０の回転力は高められる。このようにして取り出したエネルギーは再び物質物体内に取り込められる。
【００５７】
図６乃至図９に基づいて、別な実施例の形態を説明する。図６は上円形鋼板３１の底部へ凹部４５を設けてそれにＬ字形の板バネ４６，４７を合わせ２箇所ボルト４８，ナット４９で固定し、一方のＬ字形の板バネ４７の端部を上円形鋼板３１の凹部４５の側壁へ２箇所ボルトと合体したナット５０で固定したもの、或いは板バネ４７の上部を上円形鋼板３１の小さい凹部を設けて嵌合したものである。
【００５８】
図７はＬ字形板バネ４６，４７を２枚合わせた板バネの中央部へ対の一方の板バネ４７にも平行した直角の偶力を同時にかける。そのとき上円形鋼板３１は動かないように制動装置４４が働いている。それと同時に一方の端の板バネ４６の表面上にも図のように直角で平行した偶力の外力をもう一方の対にも同時にかけることにする。
【００５９】
図８は板バネ４６へ対の垂直方向の外力がかけられてその板バネ４６の底部へバネ１１ｂの表面上に固定されているキャスター１２の鋼球ボール１３を当接させて上円形鋼板３１と板バネ４６とキャスター１２，板バネ１１ｂが平行水平になるように、あらかじめそれぞれのバネの弾力が釣り合うように計算して製造し組み立てたものである。制動装置４４が働いていなかったときには、キャスター１２を通してバネ４６の底面に対して垂直の外力をかけるとキャスター１２上にある鋼球ボール１３はころがって上円形鋼板３１は図面の左方向へ偶力で押されて回転移動する。その回転方向は下円形鋼板４１の表面上に設けた２箇所の偶力受止突起１５或いはスプリング１の下部全体を嵌合した下円形鋼板４１の凹部で受ける回転原動力の偶力方向と同一である。制動装置のブレーキは最始動時まで継続してかけ続けられているので、バネ１１ｂとバネ４６を固定すると図７のようになる。バネ１１ａにある長孔１１ｃと同じ長孔をバネ１１ｂとバネ４６にもそれぞれのスプリングバネの弾性体が持つ弾力の微調節平行水平で釣り合わせるために設ける。
【００６０】
図８は完成品の上部分の拡大図である。外力を継続してかけられているところのバネ１１ｂと同様に外力を継続してかけられているバネ４６が持っている弾性力が上下円形鋼板３１，４１と平行し水平状態で釣り合ったときにボルト５１，ナット５２で固定する。バネ１１ａにある長孔１１ｃと同じ長孔をバネ１１ｂとバネ４６にもそれぞれのスプリングバネの弾性体が持つ弾力の微調節平行水平で釣り合わ

せるために設ける。
【０　０　６　１】
図９のようにバネ１１ｂとバネ４６の間にキャスター１２鋼球ボール１３があると回転動力軸３０，４０の回転力を止める制御装置のブレーキが働いたとき回転動力軸３０，４０に固定されている上下円形鋼板３１，４１は止まるが、スプリング１とバネ１１を固定して一体化したものは慣性運動によって偶力受止突起１５或いは凹部に嵌合されたスプリング１を残し、それより上部分は前方左方へ押されて回転し続けようとする。
【０　０　６　２】
しかし、ブレーキが働いたときはスプリング１はその上部で固定されているバネ１１中間部材１０を含め一体化されたものは重量と、又外力で曲げられているスプリング１とバネ１１バネ４６，４７の反力弾力があるためにブレーキがかかるとスプリング１の両端上部分は初めの外力を受けたときの反対方向に偶力の回転力を受けてその上下にある円形鋼板３１，４１に対して強力な圧縮圧力が制動装置４４のブレーキをかけると同時にかかり、その摩擦力で上下円形鋼板３１，４１と上下回転動力軸３０，４０と一体化して回転する。しかし、ブレーキをかけたときの慣性による不安定さは否めないので、図１０にあるように回転動力軸装置の最始動時にスプリング１，バネ１１，バネ４６，４７すべての弾性体が持っている弾力が調和して釣り合ったときに連結する。
【０　０　６　３】
そのとき唯一全くどこにも固定されていない部分は、上円形鋼板３１の下方に設けた上方が塞がった凹部に嵌合したバネ４７の上部分と下円形鋼板４１上にある偶力受止突起１５部分或いは凹部に嵌合されているスプリング１だけである。図１０，図１１共に外力をかけたときのスプリングとバネの作動説明図である。図１０はスプリング１と固着された中間部材１０ａ，１０ｂのＬ字形に曲げた角度は１３５度である。図１２，図１３共に、Ｙ＝１３５度からＷ＝９０度の間で曲げたものの完成品である。図１１はＷ＝９０度で曲げてあるもので完成品の一つの作動説明図である。図１２，図１３共に、Ｙ＝１３５度からＷ＝９０度の間で曲げたものの完成品である。
【０　０　６　４】
Ｙ＝１３５度前後からＷ＝９０度までの間で曲げたものならば回転動力軸３０，４０の回転力を有効に取り出せる。図１２，図１３共に、Ｙ＝１３５度からＷ＝９０度の間で曲げたものの完成品である。中間部材１０ｂに固定されたバネ１１ａの位置がＷ＝９０度からＹ＝１３５度で曲げられている部分から離れていると、てこの

原理でスプリング１の下部が支点となると小さい外力でも良いが小さい回転力しか生じない。曲げている部分から近い程、大きい回転力を生み出す。しかしあまりにも近いとバネ１１ｂとバネ４６が固定されず、キャスター１２鋼球ボール１３が間にある場合は、バネ４６から鋼球ボール１３が回転して戻りはずれることがある。
【００６５】
上円形鋼板３１の凹部４５の左側の側壁へ固定或いは動かないように嵌合されたバネ４７から生じる偶力Ｐｅによって外力Ｐａの水平方向成分Ｐｃの反力である偶力のＰｄは解消できる。回転動力軸３０，４０に負荷やブレーキが働けばＰｅの反力は現れるが、制動装置が機能している最中はそれで良い。
(12)【産業上の利用可能性】
【００６６】
本考案に係る回転動力装置は、自動車や家庭用発電機に適用することが出来る。
(13)【符号の説明】
【００６７】
１ 主スプリング、１ａ 腕部、１ｂ 腕部、２ 鋼鉄支柱、２ａ ネジ部、３ 位置決めナット部、４ 締付ナット、７ａ 上鋼球、７ｂ 中鋼球、７ｃ 下鋼球、８ カバー、９ 伸縮連結部材、１０ 中間部材、１０ａ 基片部、１０ｂ 腕片部、１０ｃ スタッド、１０ｄ ナット、１０ｅ 目玉具、１１ Ｌ形スプリング、１１ａ 下部分、１１ｂ 上片部、１１ｃ 長孔（下）、１１ｄ 透孔（上）、１１ｅ ネジ棒、１１ｆ ナット、１２ キャスター、１２ａ キャスターケース、１２ｂ スタッド、１２ｃ ナット、１３ 鋼球ボール、１４ 補強片、１５ 係止機構、１５ａ 偶力受止突起、１５ｂ 凹溝、１６ 遠心力止め螺筒、１７ 連接機構
３０，４０ 上下の回転動力軸、３１，４１ 上下円形鋼板、３１ａ 受皿、３１ｂ 保持具、３２，４２ 上下軸受、４３ 縦孔、４４ 制動装置、４５ 凹部、４６ 下板バネ、４７ 上板バネ、４８ 締付用ボルト、４９ 締付用ナット、５０ 捻込ボルト、５１ 固定用ボルト、５２ 固定用ナット、Ｐ 外力、Ｐａ 上方からの外力、Ｆ～Ｈ 力、Ｋ～Ｎ 力、Ｑ １５度、Ｒ ２２度、Ｓ ２３度、Ｔ ３０度、Ｕ ４５度、Ｗ ９０度、Ｙ １３５度

【図1】

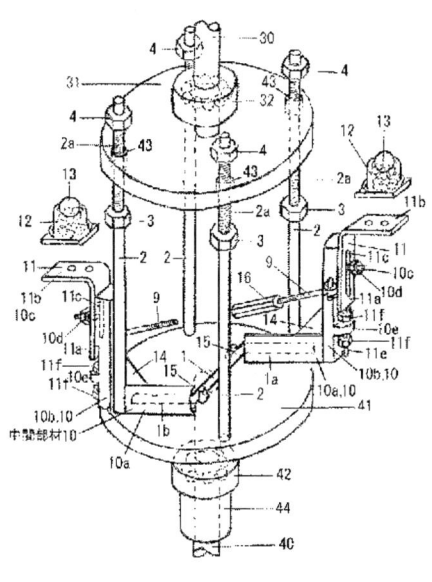

【図2】

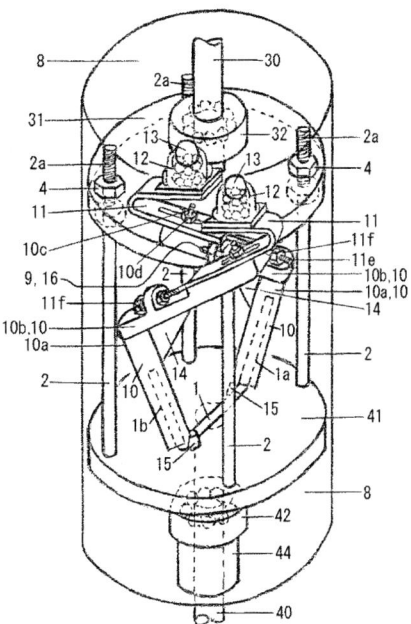

【図3】

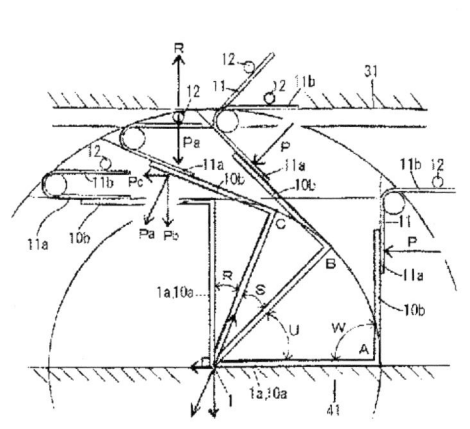

【図4】

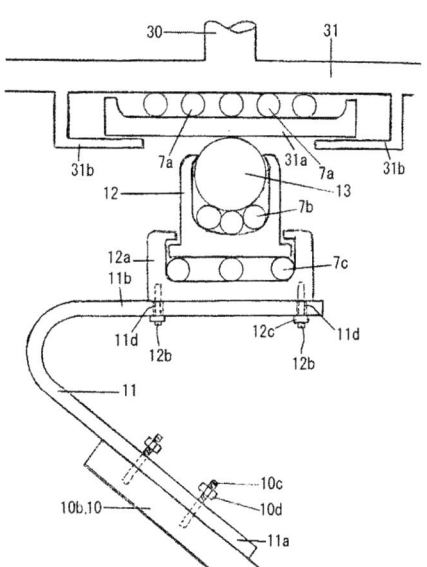

25

【図5】

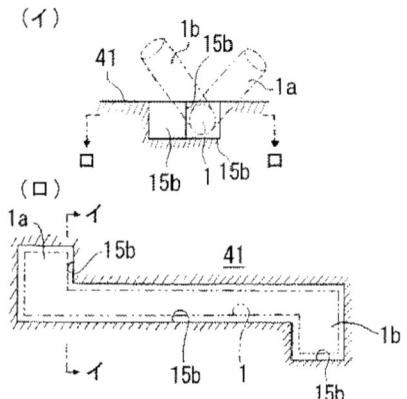

【図6】

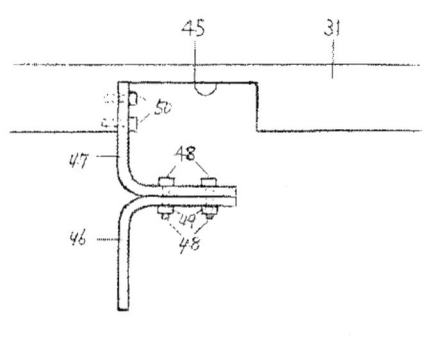

【図7】

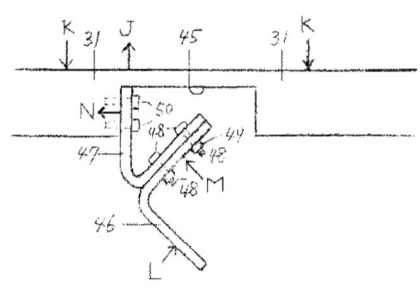

【図8】

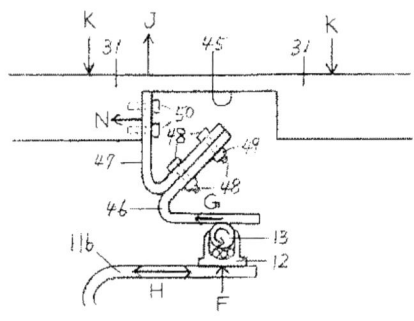

【図9】

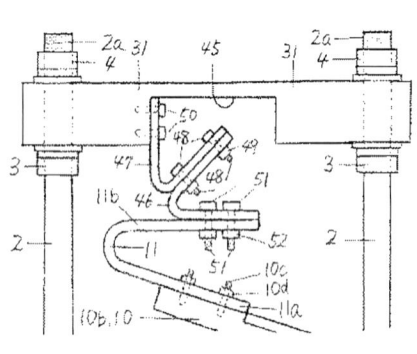

【図10】

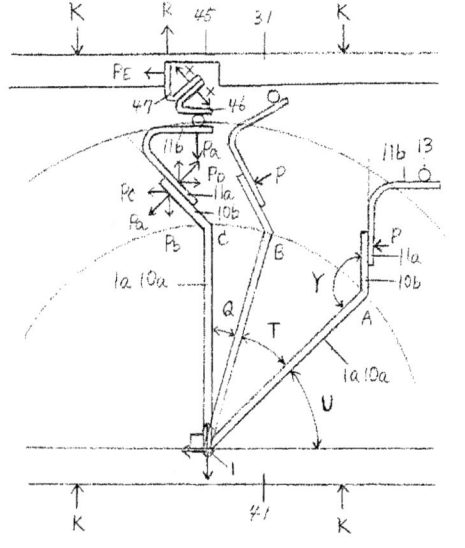

【図11】

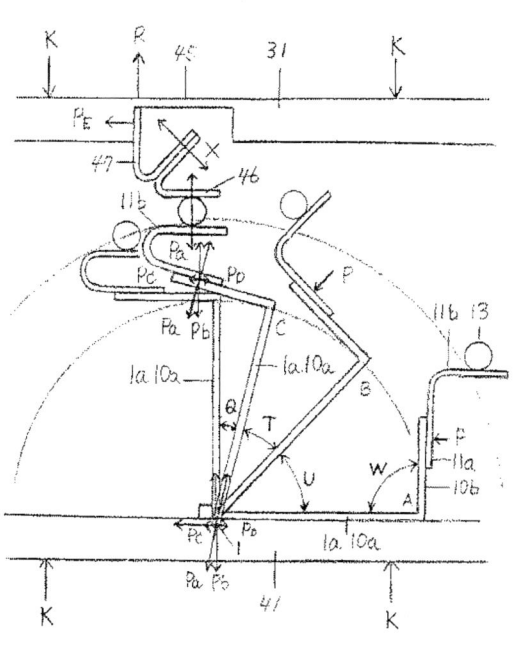

【図12】

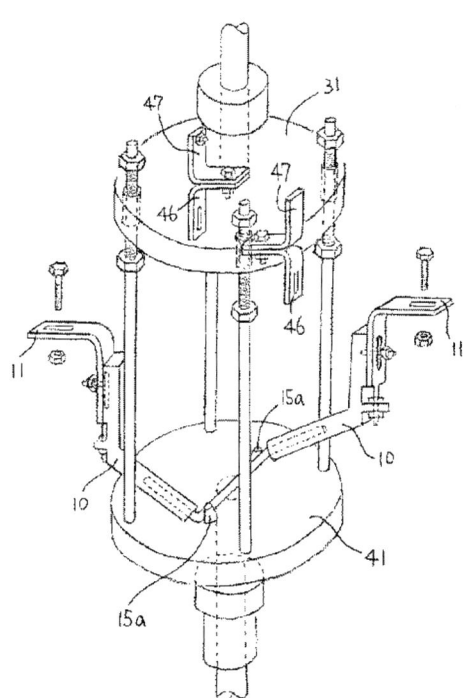

【図13】

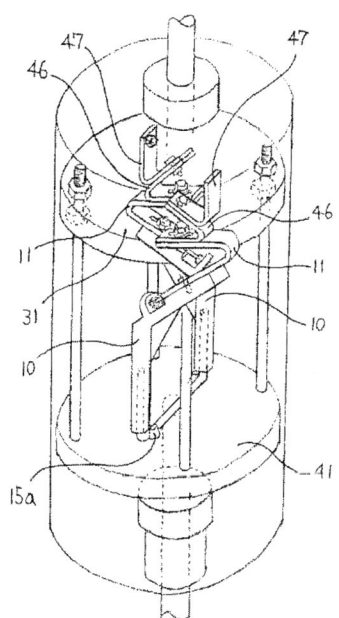

DETAILED DESCRIPTION

[Detailed explanation of the device]
[Field of the Invention]
[0001]
When external force is applied first, this device is related with the rotation power shaft equipment which can rotate for a long time, even if it removes that external force.
[Background of the Invention]
[0002]
If external force is applied first, for example, a spring type clock will be raised as equipment which rotates even if it removes the external force.
[0003]
The fine particles which constitute a substance are read with the molecule. Movement of a molecule will be restricted more strongly and a molecule will serve as a solid, if temperature falls. However, although the independent molecule exists even in such a case, among molecules, strong chemical bond power is not accepted but it is based on the van der Waals force of remarkable weak power compared with chemical bond power in many cases that the molecule is condensing. Therefore, also in such a case, it is a substance consisting of a molecule. Although a molecule does mutually the power which will be repelled if it approaches not much, when a few is left, attraction is done mutually. The more the distance between molecules becomes small, the more repulsive force will increase rapidly. Since it is not based on a chemical bond between molecules but is connected with the intermolecular force, the power is very loose.
[0004]
The more the distance between molecules becomes small, when repulsive force will increase rapidly, the more there is, but in order for the distance between two molecules to become small, a molecule and a molecule approach. Therefore, the repulsive force will be pushed and attraction will increase gradually. There, the acceleration force of the attraction of a minute distance will have occurred eternally. And two, or the attraction and the repulsive force of two or more molecules balance in a certain position, and it is only visible to a state of rest. It is the same as the thermal expansion energy and gravitation (gravity) energy of the

earth interior balance.

[0005]

There are volume elasticity started to change of volume and form elasticity started to a formal change in elasticity. The elasticity of a spring is mainly started with form elasticity. In solid form [both], form elasticity and volume elasticity start, and both become together and appear in many cases. If a stick and a board are bent, the surface of the outside of a board will be extended, and an inside surface will be shrunken, but the middle surface does not have elasticity only by bending. It is torsion deformation at the time of fixing one end of a stick or wire, and applying and twisting a couple of force to the other end, and torsional rigidity is complicated when a section is not a circle. The operation between the many fine grains in a stick and reaction are resisting.

[The outline of a device]

[Problem(s) to be Solved by the Device]

[0006]

the conventional spring -- compared with the thing of a formula, etc., it is going to provide the power plant which can rotate for a long time.

[Means for solving problem]

[0007]

It refers to Drawings and they are described.
The device of Claim 1 fixes the lower circular steel plate 41 to the upper end part of the lower rotation power shaft 40 inserted in the brake mechanism 44, The central drum section of the main spring 1 which has the both-ends arms 1a and 1b which have a crossed axes angle is locked via the lower circular steel plate 41 stopping mechanism 15, The base piece portion 10a of the two L character-like pars intermedia material 10 and 10 is fixed to the both-arms part of the main spring 1, respectively, A part for the lower part 11a of the two L form springs 11 is attached to the arm piece part 10b of the pars intermedia material 10, Construct the centrifugal-force stop connecting member 9 among the both-arms piece parts 10b and 10b, and at least three steel supports 2 are implanted in the lower circular steel plate 41, The longitudinal hole 43 of the Kami circular steel plate 31 is inserted in and fixed to the upper part of the steel support 2, the upper piece part 11b of an L form spring is arranged via the connection mechanism 17 below an upper circular steel plate, and the fixation set-up of the upper rotation

power shaft 30 is carried out on an upper circular steel plate.

[0008]

The device of Claim 2 fixes the lower circular steel plate 41 to the brake mechanism 44 and the lower-shaft carrier 42 at the upper end part of the lower rotation power shaft 40 which carried out insertion bearing, The central drum section of the main spring 1 which has the both-ends arms 1a and 1b which have a crossed axes angle is locked via the stopping mechanism 15 on the lower circular steel plate 41, Fit fixing of the base piece portion 10a of the pars intermedia material 10 and 10 which carried out the shape of two L character is carried out to the both-arms part of the main spring 1, respectively, A part for the lower part 11a of the two L form springs 11 is attached to the arm piece part 10b of the pars intermedia material 10 so that slide position regulation is possible, The centrifugal-force stop connecting member 9 which can be expanded and contracted is constructed among the both-arms piece parts 10b and 10b, Implantation fixation of at least three steel supports 2 set up near the periphery edge of the lower circular steel plate 41 is carried out, Screw the positioning nut 3 in the thread part 2a cut in the upper part of the steel support 2, and it inserts in the longitudinal hole 43 of the Kami circular steel plate 31, The nut 4 with a bundle is screwed, the upper side arm 11b of the L form spring 11 is arranged via the connection mechanism 17 below an upper circular steel plate, the fixation set-up of the upper rotation power shaft 30 is carried out on an upper circular steel plate, and ****** 32 in which axial direction justification is possible is arranged to an upper rotation power shaft.

[0009]

The device of Claim 3 constitutes the stopping mechanism 15 by two projections 15a and 15a in Claim 1 or the device of 2.

[0010]

The device of Claim 4 constitutes the stopping mechanism 15 by the concave 15b in Claim 1 or the device of 2.

[0011]

In the device of Claims 1-4, the device of Claim 5 arranges the steel ball ball 13 on the upper edge part 11b via the axle-pin rake 12, and constitutes the connection mechanism 17.

[0012]

The device of Claim 6 attaches the vertical piece of the upper board spring 47 of an L type to the concave part 45 provided on the lower surface of the upper circular steel plate 31 in Claim 1 or the device of 2, The upper side of the lower board spring 46 of the L type attached to the upper board spring 47 is attached to the horizontal piece of the upper board spring 47, and the level neighborhood of the lower board spring 46 is attached to the L form spring 11b.

[0013]

The device of Claim 7 contacts the level neighborhood of the lower board spring 46 on the steel ball ball 13 in the device of Claim 5.

[0014]

The device of Claim 8 fixes the lower circular steel plate 41 to the brake mechanism 44 and the lower-shaft carrier 42 at the upper end part of the lower rotation power shaft 40 which carried out insertion bearing, the concave part which fits in the spring 1 on the lower circular steel plate 41 -- or two couple-of-force receipt projections 15 and 15, [provide and] The central drum section of the spring 1 which has the both-ends arms 1a and 1b which have a crossed axes angle is locked to the projections 15 and 15, Fit fixing of the base piece portion 10a of the pars intermedia material 10 and 10 which carried out the shape of an L character is carried out to the arm of the spring 1, A part for the lower part 11a of the L form spring 11 is fixed to the arm piece part 10b of the pars intermedia material 10, Construct the centrifugal-force stop connecting member 9 which can be expanded and contracted between both middle components, and the steel ball balls 13 and 13 are arranged via the axle-pin rakes 12 and 12 to the upper edge part 11b of the springs 11 and 11 which attached a part for the lower part 11a to the arm piece part 10b of pars intermedia material so that slide position regulation was possible, Implantation fixation of several steel supports 2 set up near the periphery edge of the lower circular steel plate 41 is carried out, Screw the positioning nut 3 in the thread part 2a cut in the upper part of the steel support 2, and insert in the longitudinal hole 43 of the upper circular steel plate 31, and also screw the nut 4 with a bundle, start the both-arms part of a spring, make the lower surface of an upper circular steel plate abut a steel ball ball, and the fixation set-up of the upper rotation power shaft 30 is carried out on an upper circular steel plate, ****** 32 in which axial direction justification is possible is arranged to an upper rotation power shaft.

[Effect of the Device]

[0015]

The external force of a couple of force which the arms 1a and 1b of the both ends of the spring 1 are continuing applying to the surface of the springs 11 and 11 fixed to the upper part of the pars intermedia material 10 and 10 fixed in the upper part right-angled is released. That of the compression external force which is not a couple of force concerning the upper part of the springs 11 and 11 which sandwich the two axle-pin rakes 12 and 12 from the upper circular steel plate 31 which descends from ****** is a part for Shimobe of the springs 11 and 11 (from the beginning a couple of force of external force) at ******. [continue and] Although the portion 11a which is continuing being applied is pushed and tries to return to the original state with stress, it cannot return.

[0016]

When external force acts on an elastic body, the internal force (stress) which resists the external force, resists an elastic body, tries to maintain a form as it is at an elastic body, and is produced is reactionary power. stress -- composition of a molecule -- high temperature -- it comes out, and unless it changes, it continues semipermanently. Although a molecule does mutually the power which will be repelled if it approaches not much, when it separates for a while, it does the attraction between molecules mutually semipermanently. As a result, a couple of force is newly produced to a part for Shimobe of the springs 11 and 11 (upper part of the pars intermedia material 10 and 10 currently bent by L type fixed by the upper part of the arms 1a and 1b of the both ends of the spring 1).

[0017]

If external force (couple of force) other than the compression external force from the up-and-down circular steel plates 31 and 41 is eliminated as it is in the state of the Fig.2 in the state where the brake of the brake mechanism 44 is released and the vertical rotation power shafts 30 and 40 begin rotation, and C of Fig.3,In order that the lower part of the spring 1 may try to return to the original state by stress elastic force, as opposed to the same couple-of-force receipt projections 15 and 15 as the effect which produces the whole lower part of the spring 1 in the center portion concave part of the lower circular steel plate 41 which fitted in -- a couple of force -- and rotational force is semipermanently generated in the rotation power shafts 30 and 40 of the upper and lower sides which were fixed

with the up-and-down circular steel plates 31 and 41, and were unified.
[Brief Description of the Drawings]
[0018]
[Drawing 1]It is a partial exploded perspective view of the initial state of equipment.
[Drawing 2]It is a perspective view of the whole equipment.
[Drawing 3]It is an operation explanatory view of a spring.
[Drawing 4]It is a perpendicular cutting expansion front view of the essential part which deformed the part.
[Drawing 5]It is an embodiment figure when a stopping mechanism is made into a concave. (b) It is a vertical cut surface figure, and the **-** cross sectional view of (**) and (**) are plan views, and are a **-** cross sectional view of (b).
[Drawing 6]It is an initial figure of the 2nd example of a connection mechanism.
[Drawing 7]It is a middle figure of Fig.6.
[Drawing 8]It is a final drawing of Fig.6.
[Drawing 9]It is a final drawing from which Fig.6 differs.
[Drawing 10]It is an operation explanatory view of the main spring and an L form spring.
[Drawing 11]It is an operation explanatory view of the different main spring and an L type spring.
[Drawing 12]It is a partial exploded view of Fig.1.
[Drawing 13]It is a partial exploded view of Fig.2.
[The form for devising]
[0019]
It refers to a figure and it is described. In the state of distance stage A.B.C of Fig.1, Fig.2, and Fig.3,The brake mechanism 44 which controls the rotational force of the lower rotation power shaft 40 is still in the state under operation, The lower rotation power shaft 40 and the lower circular steel plate 41 have the spring 1 close by a shave on the couple-of-force receipt projections 15 and 15 or lower circular steel plate of two right and left which were fixed and were fixed to the lower circular steel plate 41, or provide the concave part in which the whole lower part of the spring 1 made [the degree which remains slightly] to carry out fitting is inserted. On the other hand, the external force P from the outside of a couple of force is applied to the both ends of the central drum section of the spring 1 in

contact with it. Though natural, while assembling this from the distance stage A to the stage of B, and also the stage of C, the external force of a couple of force from the outside must continue applying large power sequentially.

[0020]

A right angle is carried out on the surface of the base piece portion 11a which is a part for the lower part of the springs 11 and 11 fixed to the upper part of the pars intermedia material 10 and 10 which consists of the steel pipe 10a of L type simultaneously put on the arms 1a and 1b of the both ends of the spring 1 in a similar manner, and the fixed steel bar 10b, And applying the external force P of a couple of force parallel to another upper piece part 11b is continued until it will be in the state of C of Fig.3. And last becomes like Fig.2. However, don't apply the external force of the vertical direction from the outside which is parallel to the rotation power shafts 30 and 40 of the compression pressure which brings close the circular steel plates 31 and 41 of the upper and lower sides which will assist the external force P of a couple of force in the time of the distance stage B of Fig.3 in that case together at the time of said until it will be in the state of C of Fig.3. Once, it makes it such to get confused for the description for protecting. after coming to the stage of C sequentially and finishing, external force the external force P in response to the completely different external force which becomes largely and brings about the effect of the external force P of this couple of force, the same power, and power with Fig.2 and Pa of the elastic force of the springs 11a, 11b, 11a, and 11b which deformed like C of Fig.3,After applying and including in the surface of both springs 11b and 11b so that the springs 11b and 11b may be parallel to the circular steel plates 31 and 41 of the upper and lower sides of the external force of the reaction force R, Instead of the next, the up-and-down circular steel plates 31 and 41 are close brought in the position, and compressive external force is applied so that the circular steel plates 31 and 41 and the spring 11b may be parallel on the surface of the springs 11b and 11b. Therefore, new external force with the completely same effect as it is sequentially applied at the same time it eliminates external force for every stage. The distance stage A of Fig.3 will be 90 degree if it sees from vertical direction to the spring arm 1a base piece portion 10a. For B to be in the state which stood from the state of A about 45 degree in response to the external force P, for C to be in the state which occurred 23 degree from the state of B in response to the

external force P further, and what is necessary is [limitation is not carried out to the angle in the position of C, but] just an angle before and behind that.
[0021]
With the nuts 4 and 3 of the thread part 2a in a part of four steel supports 2 which fixed the lower lower circular steel plate 41 which was parallel to it in the upper circular steel plate 31 which fixed the upper rotation power shaft 30 to the center which is above Fig.1, and the upper and lower sides which fasten it. The upper circular steel plate 31 which is in between is moved below, and the up-and-down circular steel plates 31 and 41 are brought close.
[0022]
Four places of the upper circular steel plate 31 are the longitudinal hole 43 in the air, and the thread part 2a of the upper part of the four steel supports 2 passes along the inside of it.
[0023]
By fastening the nut 4 with a bundle on the upper circular steel plate 31 on both sides of the axle-pin rake 12 fixed to up to the flat surface of the upper piece part 11b of the spring 11. Suppose that it lowers so that it may be in a state parallel to the axle-pin rakes 12 and 12 who dropped gradually and were fixed above the springs 11 and 11, or level, and external force Pa of the same power as the external force in the position of the state of C of the Fig.3 from the upper part is applied. As for reaction force (however, P=Pa=R) and Pb, the vertical direction component of external force Pa and Pc of R are the horizontal components of external force Pa.
[0024]
Among the parallel up-and-down circular steel plates 31 and 41, the external force of the compression pressure of the vertical longitudinal direction different from a couple of force of the external force of the transverse direction continuously applied to the lower couple-of-force receipt projections 15 and 15 will be applied to the spring 1 and the springs 11 and 11.
[0025]
In the distance stage of A in Fig.3, although the angle of the spring 1 bent at the time of work is 90 degree, since it is a pair, it turns into 180 degree.
[0026]
If the angle of the spring 1 does not balance with the angle and strength by which

the spring 11 is bent moderate by the strength of the elastic force which it has at the time of work, it cannot pull out effectively rotational force of the up-and-down rotation power shafts 30 and 40 fixed to the up-and-down circular steel plates 31 and 41.

[0027]
Therefore, if the elasticity as an elastic body of the spring 1 is weak at the time of work, as long as the distance stage A is strong 180 degree or more at a pair, the angle of 90 degree of the stage B may be as sufficient as a line at a pair. And although the spring 11 is 90 degree similarly in the state of A and B of the Fig.3 of a preceding paragraph story where the external force from the upper part is applied, they may be the following angles more again.

[0028]
When incorporating what unified the spring 1 and the spring 11 into the up-and-down circular steel plate 31 and 41, It cannot be weak, even if it needs to be held by a horizontal state parallel to the spring 1 which was bent beforehand and manufactured, the angle of the spring 11, and the circular steel plates 31 and 41 of the upper and lower sides of the axle-pin rake 12 inserted in between in the state with optimal elastic force and either is strong.

[0029]
Begin from A like Fig.2 and Fig.3 and B,And the external force of a couple of force which it is continuing applying to the surface of the spring 11a fixed to the upper part of the pars intermedia material 10 which consists of a steel bar which the arms 1a and 1b of the both ends of the spring 1 fixed in the upper part when it changed into the state of C of a culmination right-angled is released. That of the compression external force which is not a couple of force concerning the upper part of the springs 11 and 11 which sandwich the two axle-pin rakes 12 and 12 from the upper circular steel plate 31 which descends from ****** is a part for the lower part of the spring 11 (from the beginning a couple of force of external force) at ******. [continue and] Although the portion 11a which is continuing being applied is pushed and tries to return to the original state with stress, it cannot return.

[0030]
When external force acts on an elastic body, the internal force (stress) which resists the external force, resists an elastic body, tries to maintain a form as it is

at an elastic body, and is produced is reactionary power. stress -- composition of a molecule -- high temperature -- it comes out, and unless it changes, it continues semipermanently. Although a molecule does mutually the power which will be repelled if it approaches not much, when it separates for a while, it does the attraction between molecules mutually semipermanently.

[0031]

As a result, a couple of force will newly be produced to a part for the lower part of the spring 11 (upper part of the pars intermedia material 10 currently bent by L type fixed by a part for Uwabe, the arms 1a and 1b of the both ends of the spring 1).

[0032]

If external force (couple of force) other than the compression external force from the up-and-down circular steel plates 31 and 41 is eliminated as it is in C of the Fig.3 in the state where the brake of the brake mechanism 44 is released and the vertical rotation power shafts 30 and 40 begin rotation, and the state of Fig.2,In order that the lower part of the spring 1 may try to return to the original state by stress elastic force. as opposed to the center recessed part of the circular steel plate 41 which fitted in the lower part of the couple-of-force receipt projections 15 and 15 or the spring 1 -- a couple of force -- and rotational force is semipermanently generated in the rotation power shafts 30 and 40 of the upper and lower sides which were fixed with the up-and-down circular steel plates 31 and 41, and were unified.

[0033]

The up-and-down rotation power shafts 40 and 30 are fixed to the central part of the up-and-down circular steel plates 41 and 31, and it supports by the up-and-down bearings 42 and 32.

[0034]

Although the upper part of the spring 1 tries to return with a couple of force (elastic force stress) of the same power in the direction of the origin to which the external force of a couple of force to begin was applied, The lower part of the spring 11 fixed by the external force pressure of the longitudinal direction from the upper and lower sides of the circular steel plates 31 and 41 by the upper part of the spring 1 receives simultaneously the opposite equivalent couple of force of the elastic force which tries to return in a longitudinal direction and the direction

to which the original external force was applied produced newly.
[0035]
A couple of force completely equivalent to it by narrowing the up-and-down circular steel plates 31 and 41 on both sides of the upper axle-pin rake 12 of the spring 11 passed through and fixed in between at the same time it eliminates the external force of a couple of force applied to the lower part of the spring 11 fixed to the upper part of the spring 1. A couple of force of the external force which carries out the same ******** can be applied to a part for the lower part of the springs 11 and 11.
[0036]
therefore, a couple of force generated in the lower part (couple-of-force receipt projections 15 and 15) of the spring 1 and the spring 1 -- being upper (lower part of the springs 11 and 11) -- although a couple of force to generate is the same, two upper couple of force which is come out of and generated and which faced is changed into the state of negating each other mutually.
[0037]
It tries to return to the state (A of Fig.1, A of Fig.3, B) before applying the external force of a couple of force with a couple of force of the counter direction of a couple of force applied to the both ends of the lower part of the spring 1 while the upper part of the spring 11 has fixed the axle-pin rake 12 there by elastic force. Since the axle-pin rake 12 who received the external force of the compression pressure from the up-and-down circular steel plates 31 and 41 is between **, it cannot return by any means.
It is for showing that it cannot make the portion of the spring 1 a fulcrum from the state of C to B since this has the up-and-down circular steel plates 31 and 41 even if it tries to return from C to B, and the distance stage B of Fig.3 cannot return by the circular motion although the spring 11b is written about two places by the upper part.
[0038]
Since the direction of the operation reaction of the external force of a compression pressure is parallel to the up-and-down rotation power shafts 30 and 40, it does not become the hindrance of the rotational force.
[0039]
The angle bent to L type of the pars intermedia material 10 and 10 has a

preferable angle which produces the rotational force of the most efficient rotation power shafts 30 and 40 among the up-and-down circular steel plates 31 and 41. In Fig.3, although it is 90 degree, the angle bent among 90 degree of type to L type to that around 135 degree is also good.
[0040]
Although the spring 1 is a stick spring in Fig.1 and Fig.2, it is possible also by the coiled spring or a plate-like spring.A coil spring is also possible although the spring 11 is a flat spring in Fig.1 and Fig.2.
[0041]
Since it is obstructed by the couple-of-force receipt projections 15 and 15 in the lower part unless the upper nut 4 with a bundle of the four steel supports 2 of Fig.2 is loosened since it is inserted into the up-and-down circular steel plates 31 and 41 even if it will try to return to origin for stress elasticity, if the distance stages B and C of Fig.3 are seen and, neither advance nor retreat can also be performed.
[0042]
Carrying out a deer. The up-and-down circular steel plates 31 and 41 and the main part of the rotation power shafts 30 and 40 for which stillness of the object which unified the spring 1 and the spring 11 and became independent also incorporated it with a state of rest. Since the brake mechanism 44 which controlled the rotational force of the rotation power shafts 30 and 40 from the time of beginning to apply the external force of a couple of force to the spring 1 without ** is [be / it] under operation continuously, if a brake is released, In the couple-of-force receipt projections 15 and 15, since it is continuing receiving rotational force continuously, the rotation power shafts 30 and 40 currently fixed to the up-and-down circular steel plates 31 and 41 will be in the state where it can rotate immediately.
[0043]
Since the molecule which constitutes all substance objects has generated repulsive force or attraction continuously semipermanently, acceleration force sticks and rotational speed and rotational force increase if load is not applied to the rotation power shafts 30 and 40, it must be continuously controlled by the brake mechanism 44 every regularity.
[0044]

It is required to calculate and manufacture the spring 1, the strength of the spring 11, and an angle so that the up-and-down circular steel plates 31 and 41, the upper part of the spring 11, and the axle-pin rake 12 may be in a parallel state.
[0045]
The portion which fixes a part for the lower part 11a of the arm piece part 10b of the upper part of L type of the pars intermedia material 10 and the spring 11 the long hole 11c provided to a part for the lower part 11a of the flat spring 11 as a slide form for the fine adjustment which balances the spring 1 and the spring 11, It binds tight with through and the nut 10d to the stud 10c implanted in the arm piece part 10b of the pars intermedia material 10, and the spring 11 is fixed to the pars intermedia material 10.
[0046]
Although the threaded rod 11e which carried out installation arrangement toward the bottom in the lower end edge for the lower part 11a is fixed to the eyeball implement 10e provided at the arm piece part 10b with the through nut 11f at Fig.1 and Fig.2, the two studs 10c and 10c are bound tight with through and the nuts 10d and 10d to the long hole 11c at the thing of Fig.4.
[0047]
The two studs 12b implanted in the axle-pin-rake case 12a at 11 d of bores similarly provided to the upper edge part 11b of the spring 11 are bound tight with through and the nut 12c, and the spring 11 is fixed to the axle-pin-rake case 12a.
[0048]
The receiving tray 31a which put the upper steel ball 7a on the lower surface of the upper circular steel plate 31 is arranged, and the holding fixture 31b which supports the receiving tray 31a from the bottom is fixed to the lower surface of the upper circular steel plate 31. The lower steel ball 7c is arranged in the upward opening hollow of the axle-pin-rake case 12a, disengagement prevention impossible is equipped with the axle-pin rake 12, the steel ball ball 13 is arranged via the inside steel ball 7b in the axle-pin rake's 12 upward opening hollow, and the steel ball ball 13 touches the lower surface of the upper pan 31a.
[0049]
By the powerful pressure and frictional force from the upper and lower sides received from the up-and-down circular steel plates 31 and 41, the retention of the axle-pin rake's 12 steel ball ball 13 fixed to the upper piece part 11b of the

upper part of the spring 11 is carried out without also making advance and retreat in the fixed position which touches on the surface of the back surface of the upper circular steel plate 31, and it stands it still. It unites with the up-and-down circular steel plates 31 and 41 to the concave part 45 on the circular steel plate 41, and the operation transverse direction of a couple of force which the couple-of-force receipt projections 15 and 15 receive, and the rotation power shafts 30 and 40 interlock and continuous rotation is carried out.

[0050]

As for ******** and rotation, the braking effort beyond rotational force is stopped by the rotation power shafts 30 and 40.

[0051]

Even if it receives huge frictional force, in order to adjust balance of the spring 1 and the spring 11, it can be freely rotated by the steel ball ball 13 in the inside of the axle-pin rake 12, and it does not have influence of minus on the rotational force of the circular steel plate 31 or the rotation power shafts 30 and 40 only temporarily. For that purpose, the four-fold Mie form is also required for the steel ball ball in the inside of the axle-pin rake 12.

[0052]

However, when what unified the spring 1 and the spring 11 is incorporated to the inside of the up-and-down circular steel plates 31 and 41, if the influence of minus does not arise at all to the continued rotational force to the rotation power shafts 30 and 40 which it does, frictional force is required rather -- the axle-pin rake 12 -- or, The centrifugal-force stop screw tube 16 and the elastic connecting member 9 become unnecessary, and fix the upper part of the spring 11 to the back surface of the circular steel plate 31 which is on the upper part directly. Or the rotation power shafts 30 and 40 are continued similarly, and it is made to rotate only by touching. Then, the centrifugal-force stop screw tube 16 and the elastic connecting member 9 are combined.

[0053]

Since the amount of lower part of the spring 11 fixed to the upper part of the pars intermedia material 10 is slide form, in order that the up-and-down circular steel plates 31 and 41 and the upper surface of the springs 11 and 11 may build a parallel horizontal state, the elasticity of the spring 1 and the springs 11 and 11 is fine-adjusted by a lever rule.

[0054]

The concave part of a lower circular steel plate or the couple-of-force receipt projections 15 and 15 are continued by Fig.2 at the time of the maximum start up. Just before releasing the braking effort of the brake mechanism 44, with a couple of force received, the elastic cooperation component 9 of muricoid rod form is connected with the centrifugal-force stop screw tube 16, and the both ends and the right and left springs 11 and 11 of the spring 1 are kept from opening with a centrifugal force during rotational movement. Since the balance of the spring 1, the springs 11 and 11, and the whole collapses and damage and the rotation power shafts 30 and 40 may be stopped when it opens, there is also a thing required [about two upper and lower sides]. However, regulation of balance of the spring 1 and the spring 11 ends, and at the time of restart, if it is made to abut and the back surface of the upper circular steel plate 31 and the upper surface of the spring 11b which is parallel to it are fixed with a bottle and a nut, since the centrifugal-force stop screw tube 16 and the elastic connecting member 9 are not opened with a centrifugal force during rotational movement, they are unnecessary.

[0055]

During rotational movement, since it is dangerous, the covering 8 is attached to the up-and-down bearings 32 and 42 and the brake mechanism 44, it unites with them, and this rotation power shaft equipment is installed.

[0056]

Since lower [a part of] is weak from the upper and lower sides to a pressure, it reinforces the place at which a steel pipe or the whole turned from the 135 degree order at 90 degree to L type of the pars intermedia material 10 and 10 of a steel bar with the reinforcement pieces 14 and 14 at the time of manufacture. If the spring 1 and the springs 11 and 11 use the large thing of elasticity, the rotational force of the rotation power shafts 30 and 40 will be heightened. Thus, the taken-out energy can be again incorporated in a substance object.

[0057]

The form of another working example is described based on Fig.6 thru/or Fig.9.Fig.6 provides the concave part 45 to the bottom part of the upper circular steel plate 31, sets the flat springs 46 and 47 of L type by it, and fixes them with the two-place bolt 48 and the nut 49, The upper part of the thing which fixed the end of the flat spring 47 of one L type to the side wall of the concave part 45 of the upper circular steel plate 31 with the nut 50 which united with a two-place bolt, or the flat spring 47 is provided and fitted in in the small concave part of the upper circular steel plate 31.

[0058]

Fig.7 applies simultaneously the right-angled couple of force which was parallel also to one of a pair of flat spring 47 to the central part of the flat spring with which two L type flat springs 46 and 47 were aligned. Then, the brake mechanism 44 is committing the top circular steel plate 31 so that it may not move. The external force of the couple of force which was simultaneously [with it / while] parallel by being right-angled as shown in a figure also on the surface of the flat spring 46 of an end will be simultaneously applied also to another pair.

[0059]

Fig.8 so that it may be made to abut the axle-pin rake's 12 steel ball ball 13 which the external force of a pair of vertical direction is applied to the flat spring 46, and is fixed on the surface of the spring 11b to the bottom part of the flat spring 46 and the upper circular steel plate 31, the flat spring 46, and the axle-pin rake 12 and the flat spring 11b may become level in parallel. It is calculating, manufacturing and assembling so that the elasticity of each spring may balance beforehand. While the brake mechanism 44 is not working, if vertical external force is applied to the bottom surface of the spring 46 through the axle-pin rake 12, the steel ball ball 13 on the axle-pin rake 12 rolls, and the upper circular steel plate 31 will be pushed leftward [of Drawings] with a couple of force, and it will rotate it. The hand of cut is the same as that of the couple-of-force direction of the rotation driving force which receives the whole lower part of two couple-of-force receipt projections 15 or the spring 1 provided on the surface of the lower circular steel plate 41 in the concave part of the lower circular steel plate 41 which fitted in. Since it is continuing be continued till the maximum start up, the brake of brake mechanism will become like Fig.7, if the spring 11b and the spring 46 are fixed. It provides in order to balance the same long hole as the long hole 11c in the spring 11a with the fine adjustment Taira tub bath common of elasticity which the elastic body of each spring has also in the spring 11b and the spring 46.

[0060]

Fig.8 is an enlarged drawing of the upper part of finished goods. When the elastic force which the spring 46 which continues external force like the spring 11b which continues external force and is applied, and is applied has balances by a horizontal state in parallel with the up-and-down circular steel plates 31 and 41, it fixes with the bolt 51 and the nut 52. It provides in order to balance the same long hole as the long hole 11c in the spring 11a with the fine adjustment Taira tub bath common of elasticity which the elastic body of each spring has also in the spring 11b and the spring 46.

[0061]
When the axle-pin-rake 12 steel-ball ball 13 was between the spring 11b and the spring 46 like Fig.9 and the brake of the control device which stops the rotational force of the rotation power shafts 30 and 40 works, the up-and-down circular steel plates 31 and 41 currently fixed to the rotation power shafts 30 and 40 stop, but. What fixed the spring 1 and the spring 11 and was unified leaves the spring 1 in which fitting was carried out to the couple-of-force receipt projection 15 or the concave part by inertia movement, and from it, an upper part is pushed on the method of the forward left, and tends to continue rotating.

[0062]
When a brake works, what was unified including the spring 11 pars-intermedia material 10 currently fixed in the upper part the spring 1 However, weight, Since there is reaction force elasticity of the spring 1 and spring 11 springs 46 and 47 currently bent with external force. If the brake works, the both-ends top portion of the spring 1 will start at the same time a powerful compression pressure applies the brakes of the brake mechanism 44 to a counter direction when the first external force is received to the circular steel plates 31 and 41 which are in the upper and lower sides in response to the rotational force of a couple of force, By the frictional force, it unites with the up-and-down circular steel plates 31 and 41 and the vertical rotation power shafts 30 and 40, and rotates. however, since the instability by inertia when brakes are applied cannot be denied, it is shown in Fig.10 -- as -- the time of the maximum start up of rotation power shaft equipment -- the spring 1, the spring 11, and the springs 46 and 47 -- when the elasticity which all the elastic body has harmonized and balances, it connects.

[0063]
The portion which is not then fixed anywhere at all uniquely is only the spring 1 by

which fitting is carried out to the upper part of the spring 47 which fitted into the concave part where the upper part which the upper circular steel plate 31 provided caudad was closed, couple-of-force receipt projection 15 portion on the lower circular steel plate 41, or the concave part. It is an operation explanatory view of a spring when Fig.10 and Fig.11 apply external force, and a spring.The angle which Fig.10 bent to L type of the spring 1 and the fixed pars intermedia material 10a and 10b is 135 degree. They are finished goods of what bent Fig.12 and Fig.13 among Y= 135 to W= 90 degree.Fig.11 is bent at W= 90 degree and is one operation explanatory view of finished goods. They are finished goods of what bent Fig.12 and Fig.13 among Y= 135 to W= 90 degree.

[0064]

If bent from the Y= 135 degree order before W= 90 degree, the rotational force of the rotation power shafts 30 and 40 can be taken out effectively. They are finished goods of what bent Fig.12 and Fig.13 among Y= 135 to W= 90 degree.If the position of the spring 11a fixed to the pars intermedia material 10b is separated from the portion currently bent at Y= 135 degree from W= 90 degree, when the lower part of the spring 1 serves as a fulcrum by a lever rule, small external force may be sufficient, but only small rotational force will be produced. From the portion currently bent, large rotational force is produced, so that it is near. However, when too near, when the spring 11b and the spring 46 are not fixed but the axle-pin-rake 12 steel-ball ball 13 is in between, the steel ball ball 13 may rotate from the spring 46, and return may deviate.

[0065]

Pd of a couple of force which is the reaction force of the horizontal component Pc of external force Pa is cancelable with the couple of force Pe produced from the spring 47 by which fitting was carried out so that it might not fix or move to the side wall on the left-hand side of the concave part 45 of the upper circular steel plate 31. If load and a brake work to the rotation power shafts 30 and 40, the reaction force of Pe is good while brake mechanism is functioning, although it appears.

[Industrial applicability]

[0066]

The rotational motion force apparatus concerning this design is applicable to an automobile or a home generator.

[Explanations of letters or numerals]

[0067]

1 The main spring

1a Arm

1b Arm

2 Steel support

2a Thread part

3 Positioning nut part

4 A nut with a bundle

7a Upper steel ball

7b Inside steel ball

7c Lower steel ball

8 Covering

9 Elastic connecting member

10 Pars intermedia material

10a Base piece portion

10b Arm piece part

10c Stud

10 d Nut

10e Eyeball implement

11 L form spring

11a A part for the lower part

11b Upper piece part

11c Long hole (below)

11 d Bore (above)

11e Threaded rod

11 f Nut

12 Axle-pin rake

12a Axle-pin-rake case

12b Stud

12c Nut

13 Steel ball ball

14 Reinforcement piece

15 Stopping mechanism

15a Couple-of-force receipt projection

15b Concave

16 Centrifugal-force stop screw tube

17 Connection mechanism

30 and 40 Up-and-down rotation power shaft

31 and 41 Up-and-down circular steel plate

31a Receiving tray

31b Holding fixture

32 and 42 Normal-axis carrier

43 Longitudinal hole

44 Brake mechanism

45 Concave part

46 Lower board spring

47 Upper board spring

48 ** bolt with a bundle

49 ** nut with a bundle

50 **** bolt

51 The bolt for fixing

52 The nut for fixing

P External force

Pa External force from the upper part

F-H Power

K-N Power

Q 15 degree

R 22 degree

S 23 degree

T 30 degree

U 45 degree

W 90 degree

Y 135 degree

永久運動 回転動力軸装置の使用方法

定価（本体1,000円＋税）

２０１２年（平成２４年）１１月５日発行

No. OKU-015

発行所　発明開発連合会®
東京都渋谷区渋谷 2-2-13
電話 03-3498-0751㈹
発行人　ましば寿一

Printed in Japan
著者　奥原順応 ©

本書の一部または全部を無断で複写、複製、転載、データーファイル化することを禁じています。
It forbids a copy, a duplicate, reproduction, and forming a data file for some or all of this book without notice.